AI SOLUTIONS FOR CLIMATE CHANGE AND SUSTAINABLE LIVING:

I0436415

EXPLORING THE ROLE OF ARTIFICIAL INTELLIGENCE (AI) IN ADDRESSING ENVIRONMENTAL CHALLENGES

BY
HENRY E. PARKINS

COPYRIGHT PAGE

TABLE OF CONTENTS

5

INTRODUCTION

In an era defined by unprecedented environmental change and the urgent need for sustainable solutions, the imperative to address climate change and promote sustainable living has never been more pressing. The consequences of inaction are far-reaching, impacting ecosystems, economies, and the well-being of current and future generations. Fortunately, amid these challenges, a powerful ally has emerged: Artificial Intelligence (AI). This book, "AI Solutions for Climate Change and Sustainable Living: Exploring the Role of Artificial Intelligence in Addressing Environmental Challenges," delves into the intersection of AI and environmental stewardship, offering insights into how AI can serve as a transformative force in the fight against climate change and the pursuit of sustainable living.

Importance of Addressing Climate Change and Sustainable Living

The global community faces a pivotal moment in history, as the effects of climate change manifest in increasingly severe weather events, rising sea levels, and ecosystem disruptions. Sustainable living has become a paramount consideration, encompassing resource conservation, renewable energy adoption, and the preservation of biodiversity. The imperative to address these challenges is not only a moral obligation but also a strategic necessity for ensuring the resilience and prosperity of our societies and ecosystems.

The Emergence of AI as a Potential Solution

As technological innovation continues to redefine human capabilities, AI has emerged as a potent tool for addressing complex societal and environmental issues. With its ability to process vast datasets, optimize resource allocation, and simulate complex systems, AI offers a promising avenue for developing innovative solutions to mitigate the impacts of climate change and promote sustainable living. From precision environmental monitoring to the optimization of renewable energy systems, AI has the potential to revolutionize our approach to environmental stewardship.

Overview of the Book's Structure and Objectives

This book is structured to provide a comprehensive exploration of the role of AI in addressing environmental challenges. It is designed to offer a multi-faceted understanding of AI's potential applications and impacts in the context of climate change and sustainable living. Through a combination of foundational knowledge, practical case studies, ethical considerations, and future trends, the book aims to equip readers with an informed perspective on the transformative role of AI in environmental conservation and sustainability. By illuminating the capabilities of AI and its potential to drive positive change, this book seeks to inspire and inform a wide range of stakeholders, from policymakers and researchers to technology enthusiasts and environmental advocates.

CHAPTER 1

UNDERSTANDING CLIMATE CHANGE AND ENVIRONMENTAL CHALLENGES

Overview of Climate Change: Causes, Impacts, and Current Challenges

Climate change, a complex phenomenon driven primarily by human activities, has emerged as one of the most critical challenges of our time. The release of greenhouse gases, such as carbon dioxide and methane, intensifies the greenhouse effect, leading to global warming. Human

13

activities like burning fossil fuels, deforestation, and industrial processes contribute significantly to this issue. As a result, Earth's climate is undergoing unprecedented changes, including rising temperatures, melting ice caps, and shifts in weather patterns.

The impacts of climate change are multifaceted, affecting ecosystems, weather events, and human societies. Rising sea levels threaten coastal communities, extreme weather events become more frequent and severe, and ecosystems face disruptions that can lead to biodiversity loss. The urgency to address climate change is underscored by its far-reaching consequences, affecting not only the environment but also human health, agriculture, and economies on a global scale.

Environmental Issues Affecting Sustainable Living

Beyond climate change, various environmental issues pose significant challenges to achieving sustainable living.

14

Deforestation, pollution, loss of biodiversity, and depletion of natural resources all contribute to the degradation of ecosystems. These issues not only compromise the planet's health but also jeopardize the well-being of current and future generations.

The interconnectedness of these environmental problems amplifies their impact, creating a web of challenges that demand comprehensive solutions. Air and water pollution, for instance, have direct consequences on human health, and the loss of biodiversity disrupts ecosystems' ability to provide essential services like pollination and water purification.

The Need for Innovative Solutions to Mitigate Climate Change

As the urgency of addressing climate change becomes increasingly apparent, the quest for innovative solutions takes center stage. Recognizing the limitations of traditional approaches, there is a growing

15

emphasis on harnessing the potential of artificial intelligence (AI) to tackle environmental challenges.

AI offers a unique set of tools and capabilities that can enhance our understanding of climate patterns, optimize resource management, and facilitate the development of sustainable technologies. From predictive modeling to optimizing energy consumption, AI presents a promising avenue for mitigating climate change and promoting sustainable living.

CHAPTER 2

FOUNDATIONS

OF ARTIFICIAL

INTELLIGENCE

Explanation of AI and Its Various Applications

Artificial Intelligence (AI) represents the integration of advanced computing systems that emulate human intelligence, enabling machines to learn, reason, and make decisions. The applications of AI are vast and diverse, ranging from natural language processing and image recognition to complex problem-solving. In the context of addressing environmental challenges, AI serves as a powerful tool to enhance our understanding of climate patterns, optimize resource management, and develop innovative solutions for sustainable living.

17

Types of AI Relevant to Addressing Environmental Challenges

1. Machine Learning (ML):

ML algorithms enable systems to learn from data, identifying patterns and making predictions without explicit programming. In the environmental realm, ML is instrumental in analyzing large datasets related to climate patterns, ecosystem dynamics, and pollution levels.

2. Natural Language Processing (NLP): NLP allows machines to comprehend and generate human language, facilitating communication between AI systems and users. In the context of sustainability, NLP can assist in extracting valuable insights from textual data, policy documents, and scientific literature.

3. Computer Vision: Computer vision empowers machines to interpret and

understand visual information from the world. In environmental applications, it can be used for monitoring deforestation, assessing the health of ecosystems through satellite imagery, and identifying pollution sources.

4. Optimization Algorithms:

AI-driven optimization algorithms play a crucial role in maximizing resource efficiency. From energy consumption to waste management, these algorithms help in designing sustainable practices by minimizing environmental impact.

Case Studies Demonstrating AI's Impact on Sustainable Practices

1. Climate Prediction and Adaptation:

AI-based climate models, such as those utilizing neural networks, have demonstrated improved accuracy in

19

predicting climate trends. These models contribute to enhanced understanding and preparedness for extreme weather events, assisting communities in adapting to changing climatic conditions.

2. Smart Grids for Energy Efficiency:

AI-driven smart grids optimize energy distribution by predicting demand patterns, reducing wastage, and integrating renewable energy sources effectively. This not only enhances energy efficiency but also promotes the transition towards sustainable and renewable energy.

3. Biodiversity Conservation:

AI plays a vital role in monitoring and preserving biodiversity. Through advanced image recognition and sensor technologies, AI helps track endangered species, identify poaching activities, and monitor changes in ecosystems, enabling more effective conservation efforts.

4. Circular Economy Practices:

AI supports the implementation of circular economy principles by optimizing recycling processes, reducing waste, and promoting the reuse of materials. This fosters sustainable consumption patterns and minimizes the environmental impact of production and consumption cycles.

21

CHAPTER 3

AI APPLICATIONS FOR CLIMATE CHANGE MITIGATION

AI-Driven Renewable Energy Solutions

Artificial Intelligence (AI) emerges as a cornerstone in the pursuit of sustainable and renewable energy solutions. Through advanced data analytics and machine learning algorithms, AI optimizes the generation, distribution, and consumption of renewable energy. Smart grids, powered by AI, enhance the integration of solar and wind power into existing energy infrastructure, ensuring a reliable and efficient transition towards cleaner energy sources. From predicting renewable energy output to optimizing energy storage, AI-driven solutions play a pivotal role in harnessing the full potential of renewable energy and mitigating the impact of climate change.

AI's Role in Climate Modeling and Prediction

Accurate climate modeling and prediction are paramount in understanding and addressing the complexities of climate change. AI, particularly machine learning models, significantly improves the precision and reliability of climate predictions. By analyzing vast datasets, including historical climate records and real-time observations, AI algorithms can identify patterns and make predictions about future climate trends. This capability empowers policymakers, researchers, and communities to make informed decisions and implement adaptive strategies in the face of changing climatic conditions. AI's contribution to climate modeling enhances our ability to mitigate and adapt to the impacts of global warming.

AI for Optimizing Resource Management and Conservation

Efficient resource management and conservation are essential components of sustainable living and climate change mitigation. AI-driven optimization algorithms excel in managing scarce resources, minimizing waste, and promoting conservation practices. In agriculture, AI helps optimize irrigation systems, predict crop yields, and reduce the use of harmful pesticides, contributing to sustainable food production. Furthermore, AI assists in monitoring and managing water resources, preventing deforestation, and enhancing biodiversity conservation through intelligent systems that identify and address environmental threats. By integrating AI into resource management strategies, we can achieve a more sustainable balance between human activities and the natural environment.

CHAPTER 4

AI SOLUTIONS

FOR SUSTAINABLE LIVING

Smart Cities and AI-Driven Urban Planning

The evolution of smart cities represents a paradigm shift in urban living, with Artificial Intelligence (AI) playing a pivotal role in optimizing urban planning and resource management. Through AI-driven solutions, cities can enhance efficiency, reduce environmental impact, and improve overall quality of life. Intelligent traffic management systems, powered by AI algorithms, mitigate congestion and reduce emissions by optimizing traffic flow. Additionally, AI facilitates predictive maintenance of infrastructure, waste management, and energy consumption,

25

contributing to the creation of resilient and sustainable urban environments. The integration of AI in urban planning lays the foundation for cities that are not only technologically advanced but also environmentally conscious and sustainable.

AI Applications in Agriculture and Food Sustainability

AI applications in agriculture revolutionize traditional farming practices, promoting sustainability and addressing food security challenges. Precision agriculture, enabled by AI, involves the use of sensors, drones, and machine learning algorithms to optimize crop yields, reduce resource inputs, and minimize environmental impact. AI-driven monitoring systems can detect crop diseases, assess soil health, and provide real-time insights for more efficient and sustainable farming practices. From farm to table, AI supports supply chain optimization, reducing food waste and promoting sustainable consumption patterns. The intersection of AI and

26

agriculture holds the promise of ensuring a more sustainable and resilient global food system.

AI Technologies for Managing Waste and Promoting Circular Economies

The management of waste is a critical aspect of sustainable living, and AI technologies offer innovative solutions to address this challenge. Smart waste management systems, utilizing sensors and AI algorithms, optimize waste collection routes, reduce landfill usage, and enhance recycling processes. AI plays a key role in identifying recyclable materials, sorting waste efficiently, and promoting circular economies by encouraging the reuse of resources. Through advanced analytics and machine learning, AI facilitates a more sustainable approach to waste management, minimizing environmental impact and contributing to the transition from a linear to a circular economy.

Steps on Solving Climate Change Problems in the Next 100 Years

In the face of an escalating climate crisis, the need for comprehensive and forward-thinking strategies to tackle climate change over the next century has never been more urgent. This chapter explores a visionary roadmap, outlining the essential steps that must be taken to address climate change on a global scale, with a particular focus on the transformative role of Artificial Intelligence (AI).

1. Understanding the Scale of the Challenge:

To effectively address climate change over the next century, it is imperative to first comprehend the gravity of the challenge. This section provides an in-depth analysis of current environmental trends, the impact of human activities on the planet, and the potential consequences if decisive action is not taken. Leveraging AI technologies, we can create sophisticated models to assess climate scenarios, enabling us to formulate adaptive and resilient strategies.

2. International Collaboration and Policy Frameworks:

Global challenges demand global solutions. This subsection delves into the necessity of fostering international collaboration and the development of robust policy frameworks. By utilizing AI for data analysis, policy formulation, and diplomatic negotiations, nations can work together to set ambitious targets and enforce regulations that transcend borders, laying the groundwork for a unified response to climate change.

3. Advancements in Clean Energy Technologies:

Central to mitigating climate change is the transition to clean and renewable energy sources. This section explores the potential of AI in advancing solar, wind, and other green technologies. From optimizing energy production to enhancing the efficiency of energy storage, AI can accelerate the development and deployment of sustainable energy solutions crucial for reducing carbon emissions.

4. Reforestation and Ecosystem Restoration:

Nature-based solutions play a pivotal role in climate change mitigation. This part of the chapter examines the importance of reforestation, ecosystem restoration, and biodiversity conservation. AI applications, such as satellite monitoring and machine learning algorithms, can facilitate large-scale ecosystem management, ensuring the health of vital ecosystems for carbon sequestration and biodiversity preservation.

5. Technological Innovation and Sustainable Agriculture:

Agriculture is both a contributor to and victim of climate change. This section explores how AI can revolutionize sustainable agriculture practices. From precision farming to data-driven decision-making, AI technologies can optimize resource use, reduce environmental impact, and enhance food security in the face of a changing climate.

6. Building Climate Resilience:

Adapting to the inevitable impacts of climate change is a critical aspect of the solution. This subsection delves into the role of AI in developing climate-resilient infrastructure, designing smart cities, and implementing strategies that enhance the ability of communities to withstand the challenges posed by a changing climate.

Individual Input in Solving Climate Change Problems

In the quest for sustainable living and climate change mitigation, the power of individual actions cannot be overstated. This chapter explores the pivotal role individual's play in addressing environmental challenges, examining how personal choices and behaviors contribute to the larger effort to combat climate change. Through the lens of Artificial Intelligence (AI), we explore how

31

technology can empower individuals to make informed decisions and lead environmentally conscious lives.

1. Environmental Awareness and Education:

Central to individual involvement is a deep understanding of the environmental issues at hand. This section emphasizes the importance of raising awareness and providing education on climate change. AI-driven educational platforms can deliver personalized content, fostering a sense of environmental responsibility and empowering individuals to make informed decisions about their daily lives.

2. Sustainable Consumption Habits:

Consumer choices have a direct impact on the environment. This subsection explores how individuals can contribute to climate change solutions through conscious consumption. AI-powered tools, such as apps that track carbon footprints or recommend sustainable products, can assist individuals in making eco-friendly choices in their purchasing habits, thereby reducing overall environmental impact.

3. Energy Conservation and Efficiency:

Individuals play a crucial role in energy conservation and efficiency. This part of the chapter discusses how AI applications, such as smart home technology and energy management systems, can empower individuals to optimize their energy usage. By adopting energy-efficient practices and technologies, individuals can contribute to reducing overall energy consumption and lowering carbon emissions.

4. Sustainable Transportation Choices:

Transportation is a significant contributor to carbon emissions. This section explores how individuals can make sustainable choices in their transportation habits. AI-driven solutions, such as ride-sharing optimization algorithms and electric vehicle recommendation systems, can encourage and facilitate the transition to greener transportation options, reducing the environmental impact of individual mobility.

5. Waste Reduction and Recycling:

Minimizing waste and promoting recycling are vital components of sustainable living. This subsection investigates how AI can assist individuals in managing waste responsibly. AI-powered sorting technologies, waste tracking apps, and smart recycling bins can enhance recycling efforts and encourage individuals to adopt practices that contribute to a circular economy.

6. Community Engagement and Grassroots Mo

Individual actions can have a ripple effect when amplified through community engagement. This part of the chapter explores the role of AI in fostering grassroots movements and community initiatives focused on climate change. Social media analytics, online platforms, and AI-driven communication tools can help individuals connect, share knowledge, and collectively advocate for sustainable practices at the local level.

In the face of the global climate crisis, the impact of individual actions should not be underestimated. This chapter underscores the importance of personal responsibility and highlights how AI technologies can amplify and facilitate individual contributions to climate change solutions. By empowering individuals with knowledge, tools, and a sense of collective purpose, we pave the way for a sustainable and resilient future shaped by the choices of each and every person.

Companies and Industries' Input in Solving Climate Change Problems

Enterprises and industries wield immense influence in shaping the trajectory of climate change. Let's delves into the pivotal role that businesses play in addressing environmental challenges, emphasizing the transformative potential of Artificial Intelligence (AI) to drive sustainable practices. From innovative technologies to responsible policies, the corporate sector can become a driving force in the global effort to mitigate climate change.

1. Corporate Sustainability Strategies:

This section explores the adoption of sustainable business practices as a fundamental step in addressing climate change. Companies are increasingly recognizing the need to integrate environmental considerations into their operations. AI technologies can aid in the development and implementation of comprehensive sustainability strategies, ranging from supply chain optimization to waste reduction, thereby minimizing the ecological footprint of corporate activities.

2. Green Innovations and Eco-friendly Technologies:

Innovation is at the heart of sustainable development. The chapter examines how industries can contribute to climate solutions through the development and deployment of green technologies. AI, with its capacity for data-driven insights and predictive modeling, accelerates the innovation process. Case studies on AI-driven innovations, such as energy-efficient manufacturing processes and eco-friendly materials, illustrate the transformative potential of technology in reducing industrial impact on the environment.

3. Responsible Resource Management:

Companies are significant consumers of natural resources, and responsible resource management is key to sustainability. This part of the chapter discusses how AI can optimize resource utilization in industries. From intelligent supply chain management to real-time monitoring of resource consumption, AI-driven solutions enable companies to reduce waste, conserve resources, and operate more efficiently.

4. Carbon Neutrality and Offset Programs:

Achieving carbon neutrality is a crucial objective for companies committed to addressing climate change. This section explores how AI can assist industries in measuring, reducing, and offsetting their carbon emissions. AI-driven carbon accounting tools, coupled with smart algorithms, can help companies identify emission hotspots, set reduction targets, and engage in effective carbon offset programs to achieve a net-zero carbon footprint.

5. Corporate Advocacy and Policy Influence:

Corporate influence extends beyond the boardroom, shaping public discourse and policy agendas. The chapter discusses how companies can leverage their influence to advocate for strong environmental policies and industry-wide standards. AI tools, including sentiment analysis and policy modeling, can empower businesses to make informed decisions, advocate for regulatory changes, and contribute to the development of a supportive policy landscape for sustainable practices.

6. Collaborative Initiatives and Global Partnerships:

In the face of a global crisis, collaboration is paramount. This section explores how companies can contribute to climate solutions through collaborative initiatives and global partnerships. AI-driven platforms can facilitate cross-industry collaborations, knowledge sharing, and the development of joint solutions. Case studies on successful collaborations illustrate the potential for collective action in addressing climate change at a global scale.

The corporate sector, with its vast resources and influence, is positioned to be a driving force in the fight against climate change. This chapter emphasizes the transformative role of AI in enabling companies and industries to transition towards sustainable practices. By embracing innovative technologies and adopting responsible strategies, businesses can not only reduce their environmental impact but also pave the way for a more sustainable and resilient global economy.

39

Short Term Solutions

As the urgency of climate change becomes increasingly apparent, the need for immediate, impactful solutions is critical. This chapter focuses on short-term strategies to address the pressing challenges of climate change, exploring how Artificial Intelligence (AI) can play a pivotal role in implementing swift and effective measures. From rapid response systems to quick-win innovations, the integration of AI offers practical solutions to mitigate the immediate impacts of environmental crises.

1. AI-Enhanced Disaster Response and Preparedness:

Climate change is accompanied by a rise in the frequency and intensity of natural disasters. This section delves into how AI technologies can enhance disaster response and preparedness, allowing for quicker and more efficient deployment of resources. AI-driven predictive models, real-time monitoring, and autonomous systems can assist emergency responders in anticipating, managing, and mitigating the impacts of natural disasters, ensuring the safety and well-being of affected communities.

2. Rapid Deployment of Renewable Energy Solutions:

To curb greenhouse gas emissions swiftly, a transition to renewable energy sources is imperative. This subsection explores how AI can expedite the deployment of renewable energy solutions. From optimizing the integration of solar and wind energy into existing grids to identifying ideal locations for rapid installation, AI can accelerate the shift towards cleaner energy alternatives, providing immediate relief to the environment.

3. Precision Agriculture for Sustainable Food Production:

Agriculture is both a contributor to climate change and vulnerable to its effects. This part of the chapter examines how AI-powered precision agriculture can enhance food production efficiency and reduce environmental impact. Through real-time data analysis, smart sensors, and automated decision-making, AI

41

technologies can optimize resource use, minimize waste, and promote sustainable farming practices in the short term.

4. Smart Urban Planning and Green Infrastructure:

Cities are at the forefront of climate change challenges, with growing populations and increasing urbanization. This section explores how AI can contribute to smart urban planning and the development of green infrastructure. AI-driven algorithms can assist in designing sustainable and resilient urban spaces, optimizing transportation systems, and incorporating green technologies to mitigate the immediate environmental impact of urbanization.

5. Enhanced Waste Management Solutions:

Addressing waste management is a tangible and immediate way to combat climate change. This subsection discusses how AI can optimize waste collection, recycling, and disposal processes. Smart waste bins, AI-driven sorting systems, and predictive analytics can improve the efficiency of waste management operations, reducing the environmental footprint of waste disposal in the short term.

6. Climate-Responsive Policy Frameworks:

Swift policy responses are essential to address climate change effectively. This part of the chapter explores how AI can aid in the development and implementation of climate-responsive policies. AI-driven data analysis, scenario modeling, and policy simulations can inform policymakers, enabling them to design and implement short-term measures that align with climate goals and adapt to rapidly changing environmental conditions.

In the short term, urgent action is required to address the immediate challenges posed by climate change. This chapter highlights the potential of AI solutions in providing rapid, impactful responses. By leveraging technology to enhance disaster preparedness, accelerate the adoption of renewable energy, optimize agricultural practices, plan sustainable urban environments, improve waste management,

43

and inform adaptive policies, we can make significant strides towards a more sustainable and resilient future.

Long Term Solutions

While short-term solutions address immediate concerns, the battle against climate change demands a sustained, visionary approach. This chapter delves into long-term strategies for mitigating the impacts of climate change and achieving sustainable living. By harnessing the capabilities of Artificial Intelligence (AI), we explore how technology can guide transformative changes in energy systems, ecosystems, and societal structures to secure a resilient and sustainable future for generations to come.

1. AI-Driven Climate Modeling and Prediction:

Understanding the complexities of climate change is paramount for long-term planning. This section explores how AI can revolutionize climate modeling and prediction. Advanced AI algorithms can analyze vast datasets, simulate climate scenarios, and predict future trends with greater accuracy. This predictive capability

44

empowers scientists, policymakers, and communities to plan and adapt to long-term climate changes, ensuring proactive responses to environmental shifts.

2. Sustainable Energy Grids and Decentralized Systems:

Transitioning to sustainable energy sources is a cornerstone of long-term climate solutions. This subsection investigates how AI can facilitate the development of smart, decentralized energy grids. Through predictive analytics and optimization algorithms, AI can manage and balance energy distribution from diverse renewable sources, making the transition to sustainable energy more efficient and scalable over the long term.

3. AI-Integrated Biodiversity Conservation:

Preserving biodiversity is crucial for ecosystem resilience. This part of the chapter explores how AI can aid in biodiversity conservation efforts over the long term. AI-powered monitoring systems, predictive analytics, and habitat modeling can assist in identifying at-risk species, understanding ecosystem dynamics, and implementing targeted conservation strategies, contributing to the preservation of biodiversity on a global scale.

4. Circular Economy and Sustainable Consumption:

Shifting towards a circular economy is fundamental for long-term sustainability. This section discusses how AI can support the transformation to a circular economy by optimizing resource use, reducing waste, and promoting sustainable consumption. Through data-driven insights, AI can guide the design of products, materials, and consumption patterns that align with circular principles, fostering a regenerative and sustainable economic model.

5. Climate-Resilient Infrastructure and Smart Cities:

Building climate-resilient infrastructure is essential for long-term sustainability, particularly in urban areas. This subsection explores how AI can contribute to the development of climate-resilient infrastructure and smart cities. AI-driven models can inform urban planning, optimize resource allocation, and enhance the resilience of cities to climate-related challenges, fostering sustainable and adaptive urban environments over the long term.

6. Technological Innovation and Geoengineering:

Innovative technological solutions are pivotal for long-term climate resilience. This part of the chapter explores how AI can support geoengineering initiatives aimed at mitigating climate change impacts. AI-driven simulations, data analytics, and monitoring systems can assist in developing and implementing responsible geoengineering strategies, addressing long-term climate challenges in innovative and sustainable ways.

Long-term solutions require sustained commitment and visionary thinking. This chapter underscores the transformative role of AI in guiding society towards sustainable living practices over the coming decades. By integrating AI into climate modeling, energy systems,

47

biodiversity conservation, circular economies, urban planning, and innovative geoengineering, we can foster a future where technology aligns with environmental goals, ensuring a resilient, sustainable, and thriving planet for generations to come.

Prescribing the Appropriate Technology Needed to Develop in Solving Climate Change Problems

As we navigate the complexities of climate change, the role of technology, especially Artificial Intelligence (AI), becomes increasingly significant. This chapter delves into the critical task of prescribing and developing appropriate technologies that can effectively address the multifaceted challenges posed by climate change. By leveraging AI innovations and interdisciplinary approaches, we explore how technology can be tailored to meet the diverse needs of climate change mitigation and sustainable living.

1. AI-Integrated Climate Monitoring and Data Analytics:

Accurate and timely data is the foundation for informed decision-making in climate change mitigation. This section explores the development of AI-integrated climate monitoring systems that leverage sensors, satellites, and ground-based data. Through advanced analytics and machine learning algorithms, these systems can provide real-time insights, improving our understanding of climate patterns and supporting evidence-based policy and action.

2. Precision Agriculture Technologies:

Agriculture is a major contributor to both climate change and food security. This subsection delves into the development of precision agriculture technologies powered by AI. Smart sensors, drones, and data analytics enable precision farming practices, optimizing resource use, reducing environmental impact, and promoting sustainable agricultural methods to ensure long-term food security.

3. Advanced Renewable Energy Solutions:

Accelerating the transition to renewable energy sources is essential for mitigating climate change. This part of the chapter explores the development of advanced AI-driven technologies in the renewable energy sector. From smart grids to energy storage solutions, AI can enhance the efficiency, reliability, and scalability of renewable energy systems, fostering a sustainable and low-carbon energy landscape.

4. Smart Urban Planning and Sustainable Infrastructure:

Cities are at the forefront of climate change challenges, and smart urban planning is crucial for long-term sustainability. This section focuses on developing AI-driven technologies for sustainable urban infrastructure. Smart city models, incorporating AI analytics, can optimize transportation, energy usage, and resource allocation, creating resilient and environmentally conscious urban environments.

5. Blockchain for Carbon Trading and Accountability:

Ensuring accountability in carbon emissions is vital for effective climate change mitigation. This subsection explores the development of blockchain technologies for transparent and traceable carbon trading systems. Through decentralized ledgers and smart contracts, blockchain can provide a secure and verifiable mechanism for tracking emissions, incentivizing reduction efforts, and fostering a global carbon market.

6. AI-Enhanced Climate Adaptation Technologies:

Adapting to the impacts of climate change requires innovative technologies. This part of the chapter examines the development of AI-enhanced climate adaptation solutions. From predictive modeling to autonomous systems, AI technologies can assist in designing and implementing adaptive measures, safeguarding communities, ecosystems, and critical infrastructure from the effects of a changing climate.

7. Responsible AI Governance and Ethical Considerations:

As we prescribe technologies for climate change, ethical considerations and responsible AI governance are paramount. This section explores the development of frameworks to ensure the ethical deployment of AI in climate solutions. From privacy concerns to algorithmic bias, addressing these ethical considerations is crucial in building public trust and ensuring the responsible use of AI technologies in the fight against climate change.

Prescribing the appropriate technology for climate change solutions requires a holistic and interdisciplinary approach. This chapter underscores the transformative potential of AI in developing tailored technologies for climate monitoring, sustainable agriculture, renewable energy, smart urban planning, carbon trading, and climate adaptation. By

embracing ethical considerations and responsible governance, we pave the way for a future where technology becomes a powerful ally in the global effort to address climate change and promote sustainable living.

CHAPTER 5

ETHICAL AND REGULATORY CONSIDERATIONS

Ethical Implications of AI in Environmental Conservation

The integration of Artificial Intelligence (AI) into environmental conservation efforts raises profound ethical considerations. As AI systems become key decision-makers in resource management, biodiversity conservation, and climate mitigation, it is crucial to address concerns related to transparency, accountability, and unintended consequences. Ethical questions surround issues like algorithmic bias, data privacy, and the potential for socio-economic disparities arising from unequal access to AI-driven solutions. Striking a balance between the benefits of

AI in environmental sustainability and ethical considerations requires careful consideration, stakeholder engagement, and a commitment to ensuring that AI technologies serve the collective good without exacerbating existing inequalities.

Regulatory Frameworks and Policies Governing AI in Sustainable Initiatives

As the role of AI in addressing climate change and sustainable living becomes more prominent, the need for robust regulatory frameworks and policies becomes imperative. Governments and international bodies must establish clear guidelines governing the development, deployment, and oversight of AI technologies in environmental initiatives. Regulatory frameworks should address issues such as data privacy, algorithmic transparency, and accountability for the environmental impact of AI systems. Collaboration between policymakers, industry leaders, and environmental advocates is essential to create regulations that foster innovation while safeguarding against potential risks and ethical concerns associated with AI in sustainable initiatives.

Ensuring Responsible and Equitable Deployment of AI Solutions

The responsible deployment of AI solutions in the context of climate change and sustainable living requires a commitment to fairness, transparency, and inclusivity. Striving for equity in access to AI technologies ensures that the benefits are distributed across diverse communities and regions. Additionally, developers and organizations must prioritize transparency in AI decision-making processes, enabling stakeholders to understand and question the outcomes generated by these systems. Ensuring accountability for the environmental impact of AI solutions is essential, with organizations taking

responsibility for mitigating potential negative consequences. Ethical guidelines and codes of conduct should be established within the AI community to guide responsible development and deployment practices, fostering a culture of accountability and ethical responsibility.

CHAPTER 6

INNOVATIONS AND CASE STUDIES

Showcasing Cutting-Edge AI Technologies for Climate Change and Sustainability

The rapid evolution of Artificial Intelligence (AI) technologies is opening new frontiers in the battle against climate change and the pursuit of sustainable living. Cutting-edge AI solutions are leveraging advanced algorithms, machine learning, and data analytics to address environmental challenges with unprecedented precision and efficiency. From predictive modeling for climate adaptation to optimizing renewable energy systems, these technologies are at the forefront of innovation. This section explores the latest advancements in AI, shedding light on how

58

these technologies are reshaping our approach to environmental conservation and sustainable practices.

Case Studies Highlighting Successful AI Implementations in Real-World Environmental Challenges

Real-world case studies offer tangible evidence of the impact AI can have on addressing environmental challenges. Through successful implementations, organizations and communities are demonstrating the transformative power of AI in mitigating climate change and promoting sustainable living. Examples may include AI-driven solutions for precision agriculture, wildlife conservation, or waste management. These case studies showcase the tangible benefits of integrating AI into environmental initiatives, providing valuable insights into the diverse applications and potential positive outcomes.

Lessons Learned and Best Practices for Leveraging AI in Environmental Initiatives

Drawing on the experiences of organizations and initiatives that have effectively deployed AI for environmental sustainability, this section outlines key lessons learned and best practices. Understanding the challenges, successes, and pitfalls encountered in real-world scenarios is essential for refining strategies and optimizing the deployment of AI technologies. From ensuring ethical considerations to navigating regulatory frameworks, these lessons offer practical guidance for stakeholders involved in leveraging AI for climate change mitigation and sustainable living. As the field continues to evolve, these insights contribute to a collective knowledge base

that informs the responsible and impactful use of AI in environmental initiatives.

CHAPTER 7

FUTURE TRENDS AND OPPORTUNITIES

Emerging Trends in AI for Climate Change and Sustainable Living

As we move into the future, the landscape of Artificial Intelligence (AI) for climate change and sustainable living is marked by dynamic and emerging trends. Continuous advancements in machine learning, data analytics, and computational capabilities are shaping the next generation of AI solutions. Emerging trends include the integration of AI with Internet of Things (IoT) technologies for enhanced environmental monitoring, the rise of explainable AI to address transparency concerns, and the application of AI in climate adaptation strategies. The development of AI-powered digital twins for

62

modeling and simulating complex environmental systems is also gaining traction. This section explores these and other trends that are shaping the future of AI in the realm of environmental sustainability.

Opportunities for Further Research and Collaboration in This Field

The dynamic nature of AI and its intersection with climate change and sustainable living present abundant opportunities for further research and collaboration. Future endeavors could focus on refining AI models for even more accurate climate predictions, expanding the applications of AI in circular economy practices, or exploring novel ways to harness AI for biodiversity conservation. Collaborations between researchers, policymakers, industry leaders, and environmental advocates can facilitate the development of interdisciplinary solutions that address complex challenges.

63

Additionally, exploring the societal and ethical implications of AI in environmental initiatives will be crucial for building a comprehensive understanding and ensuring responsible deployment.

Envisioning a Sustainable Future Empowered by AI Solutions

Envisioning a sustainable future empowered by AI involves leveraging the potential of these technologies to create positive and lasting impacts on the environment. AI can play a central role in achieving carbon neutrality, optimizing resource usage, and fostering a global transition towards sustainable practices. As smart cities evolve, AI will contribute to creating urban environments that are resilient, energy-efficient, and environmentally conscious. The integration of AI in agriculture will continue to enhance food security while minimizing environmental impact. Furthermore, the collaboration between AI and renewable

energy technologies holds the promise of a future where clean energy sources are efficiently harnessed to meet global demands.

CHAPTER 8

CONCLUSION

Recap of Key Insights and Findings

Throughout this exploration of "AI Solutions for Climate Change and Sustainable Living," we have delved into the multifaceted role of Artificial Intelligence (AI) in addressing pressing environmental challenges. Key insights include understanding the nuances of climate change, the application of AI in diverse sectors from renewable energy to waste management, and the ethical considerations that accompany the integration of advanced technologies into environmental initiatives. Real-world case studies and innovations showcased how AI can transform these challenges into opportunities, offering a glimpse into the potential for positive change.

Call to Action for Leveraging AI in Addressing Environmental Challenges

As we stand at the intersection of technology and environmental stewardship, there is a compelling call to action. Leveraging AI for climate change and sustainable living requires collaboration among governments, businesses, researchers, and communities. We must advocate for responsible development, deployment, and regulation of AI technologies, ensuring that they align with ethical standards and contribute to equitable and sustainable outcomes. This call extends to individuals as well, encouraging everyone to be conscious consumers, adopt sustainable practices, and actively engage with AI-driven solutions that promote environmental well-being.

Closing Thoughts and a Look Ahead to the Future of AI and Sustainability

In closing, the marriage of AI and sustainability holds immense promise for the

Looking ahead, it is essential to keep an eye on emerging trends, foster interdisciplinary research, and maintain a proactive stance in addressing ethical considerations. AI is a powerful tool, and its role in creating a sustainable future is contingent on our ability to harness its potential responsibly. Through collective efforts and a commitment to environmental stewardship, we can pave the way for a harmonious coexistence between AI and the planet, ensuring a resilient and sustainable future for generations to come.

CHAPTER 9

APPENDIX

Glossary of Key Terms and Concepts

To enhance the understanding of key terms and concepts discussed throughout this book, a glossary is provided below:

1. **Artificial Intelligence (AI):** The development of computer systems that can perform tasks requiring human intelligence, such as visual perception, speech recognition, and decision-making.

2. **Climate Change:** Long-term changes in temperature, precipitation, and other atmospheric conditions on Earth, often attributed to human activities.

3. **Renewable Energy:** Energy derived from resources that are naturally replenished on a human timescale, such as sunlight, wind, rain, tides, waves, and geothermal heat.

4.   **Smart Cities:** Urban areas that utilize technology, particularly AI, to improve efficiency, sustainability, and the overall quality of life for residents.

5.   **Precision Agriculture:** The use of technology, including AI, to optimize farming practices, enhance crop yields, and minimize resource inputs.

6.   **Circular Economy:** An economic system that prioritizes the continual use of resources through recycling, reusing, and reducing waste.

Additional Resources for Further Reading

For those interested in delving deeper into the topics discussed in this book, the following resources are recommended:

1.   **Reports and Publications:**

•   **Intergovernmental Panel on Climate Change (IPCC) Reports**

•   **World Economic Forum's Reports on AI for Sustainability**

2.   **Online Platforms:**

- **AI for Earth by Microsoft**

- **Climate Change AI**

3. **Books:**

- **"The Uninhabitable Earth" by David Wallace-Wells**

- **"AI and Climate Change" by Heiko Borchert**

OTHER BOOKS BY THE AUTHOR

USINESS ETHICS AND CORPORATE SOCIALRESPONSIBILITY: CREATING A PURPOSE-DRIVEN COMPANY

SIDE HUSTLE HANDBOOK: EARNING EXTRA INCOME FOR FINANCIAL INDEPENDENCE

A GUIDE TO TRACKING CRYPTOCURRENCY WHALES: SCALE-UP YOUR CRYPTOCURRENCY PORTFOLIOS (10,000X)

BUILDING RESILIENCE AND EMOTIONAL

INTELLIGENCE

INVESTOR'S GUIDE: THE POWER OF FUNDAMENTAL

ANALYSIS OF FOREX TRADING

WHAT BILLIONAIRES DOES TO BECOME WEALTHY:

UNDERSTANDING THE GAME

THE RULE OF THE BILLIONAIRES

DISCOVER SELF DEVELOPMENT BOOKS
A PRACTICAL GUIDE TO FINANCIALSUCCESS: STEPS TO WEALTHAI WEALTH: YOU'RE GUIDE TO MAKING MONEY USING ARTIFICIAL INTELLIGENCE (AI) FROM HOM

THE POVERTY RATE IN AMERICA: THE REALITIES, IMPLICATIONS, AND SOLUTION

REDICTIVE ANALYTICS FOR CUSTOMER BEHAVIOR:

USING AI TO PREDICT CUSTOMER BEHAVIOR

TRENDS HELPS IN MAKING DATA-DRIVEN

MARKETING DECISIONS

HOW TO FIND HAPPINESS AND FULFILLMENT

HOW TO IMPROVE YOUR RELATIONSHIPS

HOW TO OVERCOME ANXIETY AND STRESS:

SUMMARY

AUGMENTED REALITY (AR) AND AI IN MARKETING:

EXPLORING HOW AI AND AR CAN CREATE

IMMERSIVE MARKETING EXPERIENCES FOR

CONSUMERS, SUCH AS VIRTUAL TRY-ON

EXPERIENCES OR INTERACTIVE PRODUCT

73

Note Page 5..........................Date................

Use this page for all writing during reading or study.

Note Page 5...........................Date................

Use this page for all writing during reading or study.

Note Page 5..........................Date................

Use this page for all writing during reading or study.

Note Page 5..........................Date...............

Use this page for all writing during reading or study.

Note Page 5..........................Date................

Use this page for all writing during reading or study.

Note Page 5...........................Date................

Use this page for all writing during reading or study.

Note Page 5...........................Date................

Use this page for all writing during reading or study.

Note Page 5...........................Date................

Use this page for all writing during reading or study.

Note Page 5...........................Date................

Use this page for all writing during reading or study.